LRC

Contents

Bags, Bowls, Buckets, and Baskets

Have you ever stopped to think about the role that bags, bowls, baskets, and other containers play in your life? Probably not! But the truth is, containers are very important to the day-to-day lives of people around the world.

We tend to forget about containers because we're so busy thinking about the things inside them. But how would we ever carry, protect, and organize the important things in our lives if we didn't have containers?

Imagine going through a day without containers. You might sit down to eat breakfast, but you wouldn't have a bowl to pour your cereal and milk into. Then what would you do? You'd probably have to eat out of your hands, which wouldn't work so well. Maybe you'd have a sandwich and some fruit you want to bring to school for lunch. But there would be no bags or lunchboxes anymore. Now what? And think of what your house would look like: The houseplants would have lost their pots. With no baskets to hold the laundry, there might be clothes all over the place. All the food in the kitchen would be sitting in little piles and puddles, some of it probably starting to go rotten and smell. What a mess!

Now do you think containers are important?

Girl carrying water jug, Guatemala

Banana-leaf bowls, Indonesia

Tlingit Indian basket,
United States

You'll find containers in every culture around the world. Not only do they come in all sorts of shapes, colors and sizes, but they have all sorts of uses, too. And no matter how different one container may look from another, they often will have a similar use.

Looking at these containers from around the world, can you name something from your culture that has a similar purpose?

Shopping at a local market,
Russia

Celebration in a rural village,
Cameroon

Wood, Clay, Straw, or Glass?

Though many of the containers we use today are made from plastic, cardboard, or glass, throughout history people have turned to the earth for the materials needed to make containers. Even as technology grows more advanced, natural products such as wood, clay, and plant fibers still provide convenient and inexpensive materials for making containers.

Probably the oldest and most widely found natural material is clay. As long ago as 10,000 B.C., people used clay—a mixture of very small rock particles and water—to make pottery. A clay pot is good for cooking because it won't burn in a fire the way a wood container or basket would. Another important property of clay is that it can hold water and even keep it cool. In hot climates, such as those in South America, large clay pots are often used to store drinking water.

Woman making clay pot, Ecuador

Fruit for sale, Indonesia Today, plastic containers are used in many parts of the world.

Grocery store, Vietnam Most of the foods people buy in stores are packaged in containers made of manufactured materials like glass, cardboard, paper, and tin.

Hand-carved wooden box, Romania In countries that have many trees, traditional containers may be made out of wood.

Ancient clay urns, Greece People have been making containers out of clay for thousands of years.

When people couldn't find clay in their natural environment, they often looked to trees and plants for help in making their containers. Not only could they carve wood to make bowls and boxes, but they could also use grasses and fibers to weave baskets. The Tarahumara people of Mexico are famous for making baskets from the leaves of the yucca plant, which is found commonly all over the southwestern United States and in Mexico.

Perhaps the most important plant material of all is the gourd. Gourds are fruits that grow on certain kind of plants. They come in all sorts of sizes and can be scooped out and dried in the sun so that what's left is a very sturdy shell, which can be used as a bowl or bottle.

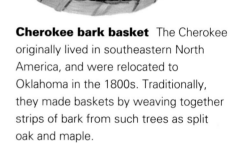

Cherokee bark basket The Cherokee originally lived in southeastern North America, and were relocated to Oklahoma in the 1800s. Traditionally, they made baskets by weaving together strips of bark from such trees as split oak and maple.

Yucca baskets, Mexico The easiest way to make a basket is to gather plant fibers that grow in your natural environment. The Tarahumara people of northern Mexico are famous for their beautiful baskets, made from the yucca and pine needles of the desert.

Unlike clay, gourds are very
lightweight, and therefore make good
containers for carrying long distances.
In Nigeria and Mali, women fill gourds
with things like milk, peanut oil,
or grains, and carry them to market
on trays balanced on their heads.

**Women buying calabashes at a local
market, Mali** In West African countries,
large gourds called *calabashes* are used
as containers. Women in Mali find these
"head gourds" the most convenient way
to transport their goods to and from the
market.

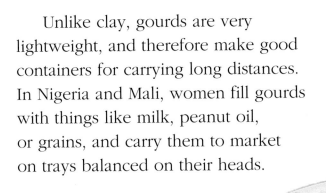

**Man selling beverage in bamboo
containers, Indonesia** Bamboo
grows plentifully in the warm, wet
countries of Southeast Asia. Bamboo
stalks make good containers for liquids
because they are extremely hard and
waterproof.

Coconut cup, Polynesia
A coconut shell cut in half and emptied
out can make an ideal drinking cup!
In Thailand, palm trees that bear
coconuts are plentiful.

People have also traditionally made containers from the skins of animals native to their environment. An Inuit hunter in Greenland might carry a pouch made of sealskin, while in Peru, farmers have used the tough shell of the armadillo to carry and protect the eggs laid by their chickens. By looking at the materials certain people use to make containers, you can begin to form some ideas about what their natural environment must be like.

Inuit reindeer-hide pouch
The Inuit people of northern Canada and Greenland live largely on tundra that has no grass, trees, or mud. Instead, they turn to native arctic animals for natural materials. This pouch is made from reindeer hide.

Armadillo-shell container, Mexico
Farmers in ancient Mexico found that the sturdy shell of the armadillo could be made into a good protective holder for corn kernels during planting season.

Onyx box, Pakistan This decorative box is carved from onyx, a type of stone found in the Asian country of Pakistan.

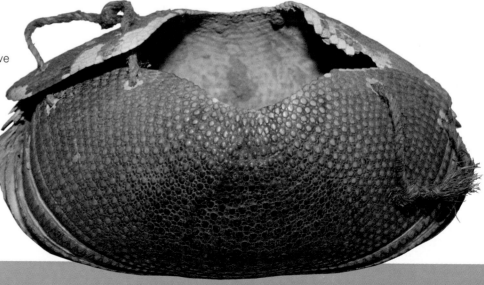

Ostrich-egg jars, Botswana The San people of Botswana in Africa have found a readily available lightweight container in which to carry water—an empty ostrich egg shell! These eggs, which are about the size of bowling balls and whose shells are very strong, can also be used as cooking pots.

Goatskin water bags, Niger The Fulani are a mainly nomadic people who live in the desert and savanna regions of West Africa. They need to carry along lots of water as they move from place to place in search of grazing land for their animal herds. These children are using a leather bucket to pour well water into goatskin water bags.

Sauk and Fox *parfleche* trunk, North America On the Great Plains of North America, where buffalo were once plentiful, Indians traditionally made stiff, durable carrying bags from *parfleche*—raw buffalo hide that has been cleaned, stretched, and dried in the sun. Why did they use rawhide instead of leather? Since their lifestyle was nomadic, they didn't have time to tan the skins.

Holding the Necessities

You've probably noticed that a lot of containers are made for holding food and water. This is because not only do we need food and drink to survive, but without containers to store, keep fresh, and protect food and water from things that might spoil it, such as heat and dirt, we could be in real trouble! In a place like Siberia, where nothing grows during the winter and grocery stores are scarce, people need to store their food in containers that will help it last through the long season. Likewise, if you were to take a long trip across the desert, you'd need to carry lots of containers full of water. In what kinds of containers are your food and drink stored in?

Honey bundled in leaves, Zaire

Gourd water jug, Kenya
Gourds are used to carry water in many parts of the world. The narrow neck of the gourd allows it to be carried on long, uphill treks from the river valleys without spilling.

Ceramic water jar, Mexico
Clay pots are especially good for storing water in hot climates, because the moisture that seeps through the clay and then evaporates helps to keep the pot nice and cool.

People eating food at an outdoor market, Thailand

Masai drinking gourds, Kenya Among the Masai, the milk and blood of cows are essential foods. Herders use long thin gourds to collect a small amount of blood from a cow's neck. (This doesn't harm the cow). The blood is then either drunk alone or mixed with milk.

Milk bottle, England In England, some people have fresh bottles of milk delivered to their very own doorstep!

Palm-frond bowl, American Samoa In Samoa, some people prepare or serve food in bowls woven out of coconut-palm fronds. This bowl will be used to carry taro root, a starchy potato-like food, to an underground oven nearby.

Carriers of Culture

Inca drinking vessel, Peru
This cup is shaped like the head of
a jaguar—an animal much respected
and feared by the ancient Inca people.

In many respects, containers are not just important to people's survival, they actually help with the survival of culture as well. As civilization becomes more advanced and we rely more and more on factories to produce the goods we need, we risk losing the cultural traditions—specific things that make one community different from another—that have been handed down through generations. Often these traditions take the form of crafts, like basket-weaving, making clay pots, or carving wooden boxes. Does anyone in your family or town have a craft that you'd like to learn?

Looking at containers from around the world can help us to appreciate the unique aspects of other cultures that have survived for many generations. For example, Chinese artists still make the delicate blue-and-white porcelain vases that their ancestors from the Ming Dynasty made 500 years ago.

**Chinese vase, Ming Dynasty
(1368-1644)** These lovely porcelain
creations have been popular in China
and other parts of the world for more
than 600 years! This ability to make
beautiful delicate dishes led to the
western term *china*, used to describe
fancy porcelain dishware.

Etched gourd, Colombia This gourd bowl from Colombia
has been etched with scenes from a village festival. Other, similar
gourds tell picture-stories about the life of ranching or farming
that many Colombians lead.

Mexican peñas These pineapple-shaped glass jars are not only pieces of art, but are also used to store *tepache,* a pineapple drink.

Pueblo pottery, southwestern United States The Pueblo people have a variety of different styles of pottery, one of which is the black-and-white pottery of Acoma Pueblo in New Mexico.

The traditional pottery made by the Pueblo people in the American Southwest is thicker and heavier than Chinese porcelain, and often painted with symbols of the animals and plants of the desert—much different than the Chinese pottery but equally as beautiful! Luckily, in both of these cultures, there are young people who are learning the craft of their older relatives and therefore carrying on tradition. Because these traditions help to demonstrate what makes a culture special, it becomes important to continue them, so some day we don't end up all the same!

***Inrō*, Japan** The *inrō*, a small case with several compartments to hold medicines, sweets, perfume, or other small items, was once a part of the traditional Japanese male costume. The *inrō* was hung by a silken cord from the *obi* (sash) of a man's kimono. Elaborately decorated *inrōs* are among the finest examples of Japanese craftsmanship.

Lacquered box, Russia Lacquerwork is an ancient, highly respected art form in many Asian countries. The process of lacquering is very elaborate and requires a good deal of training. Lacquer, which comes from the sap of the sumac tree, is painted in many shiny thin layers over a delicate wooden box. After the layers have hardened, a design or landscape may be added on top of or incised into the lacquered surface.

Yanomami Indian decorating basket, Brazil

Ndebele beaded bag, South Africa The Ndebele people are known for the brightly colored geometric designs used not only on their beaded bags and jewelry, but also on the walls of their buildings.

Meissen porcelain vase, Germany
Meissen is a type of high-quality porcelain that has been produced in Dresden, Germany, since 1710. Antique Meissen is often decorated with *deutsche Blumen* ("flowers native to Germany") taken from books of botanical illustrations.

Lacquered box, Myanmar The lacquer used in Myanmar comes from a gummy deposit created by a certain type of insect.

Inlaid box, Egypt Inlay is a decorative process in which tiny pieces are set into a surface to form a pattern.

Leather *maté* containers, Uruguay These traditional containers are meant to hold *maté*, a tea-like beverage popular in many South American countries.

Holding Position

If you think about it, many containers are made to hold the things we most value and want to protect. For instance, we keep things like jewelry, silverware, and even money tucked away in special boxes. In other instances, containers—because they give the world an outward impression—are made from precious materials and can be more valuable than their contents. In Europe and China in the 1800s, men carried decorated "snuff" boxes. Snuff is dried tobacco that's been ground into a powder. Even though the snuff wasn't worth much, it became very important for men to show their wealth by storing their snuff in very fancy boxes, which they would keep on a table or mantel as a sign of wealth and nobility.

Dowry chest, India At one time, it was common in many cultures for a woman to have a dowry chest that held linens, silver, and other objects of value, which was given to her husband once she married. The wealth of the bride's family was indicated by the size of the dowry.

Pill box, England, 1700s Wealthy European men and women often wore ornate miniature pill boxes on a chain around their waist or neck. After a meal, they might take a certain medication, and then pass around their pill boxes so others could admire them!

Fabergé egg, Russia Peter Carl Fabergé was a 19th-century Russian craftsman who made boxes decorated with precious materials such as gold, silver, jade, and gems. He was renowned for his Imperial Easter eggs, which became the delight of Russian and other royalty throughout Europe and Asia. Today collected by museums and art lovers around the globe, Fabergé's work is among some of the world's most expensive art.

Ancient cocoa bag, Mochica culture, Peru A fancy or expensive container can show off one's wealth or status in society. This puma-shaped gold bag is intended to hold cocoa beans, which were used as money in many ancient Central and South American cultures. It was probably owned by a ruler or high-ranking member of the Mochica culture.

Antique snuff box, China Snuff boxes were usually fairly small but extremely elaborate.

Pomo gift basket, North America This type of basket, made by the Pomo people of California, represents years of work, during which the basketmaker would gather the feathers of birds native to that environment, such as orioles, woodpeckers, and pheasants, and weave them into the basket. The basket is not meant to be used, but rather to be hung as a work of traditional art.

For similar reasons, men in Zaire traditionally took great pride in the wooden boxes in which they stored their shaving razors and would often show them off to their friends!

Protecting Beliefs

Besides food, tradition, and belongings, what else is valuable to a culture? Well, one thing to consider is that many cultures are formed around beliefs and rituals. Interestingly, containers have been used throughout history to protect and treasure beliefs. Whether they're used in ceremonies, or they're decorated to depict a people's beliefs, different containers can express just how important beliefs are to culture. In the Philippines, people are often buried with the best pottery they've ever made, so the gods will know how accomplished they are!

Shaker boxes, United States
The plain, simple design of boxes made by the Shaker people, an American religious society, reflects the importance they placed on modesty. Because neatness was also a way of showing faith in God, Shakers, who lived in communal houses, relied heavily on containers to keep their few possessions tidy.

Celadon pottery, China
These plates were rumored to have magical powers. Celadon pottery was supposed to break or change color when food that had been poisoned was placed on it!

Kola nut box, Nigeria This brass box is made to store kola nuts, which are chewed as part of a sacred ritual in parts of Africa.

Japanese tea ceremony *Chanoyu,* the ceremonial art of making tea, is a ritual common in Japan. The delicate taste of the tea and the simplicity of the ceremony is meant to help its participants achieve serenity and an understanding of beauty.

Ceremonial vessel, India This special copper vessel is used to hold water from the Ganges River, which is considered sacred by people of the Hindu faith.

In the 1630s, when the Puritans made the long journey by ship to America, they carried their Bibles in beautifully carved wooden boxes that represented the importance of bringing their faith carefully to a new land.

Temple offerings, Bali, Indonesia During certain religious festivals, Balinese women walk to the local temple carrying elaborate offerings of fruit, piled high on modern-day basins imported from factories in Hong Kong.

Ceramic burial urn, Guatemala, c. 800
The ancient Mayans cremated their dead and stored the ashes in ceramic urns. Some of these ancient burial vessels are decorated with figures meant to signify the spirits of the deceased.

Gilded censers, China A censer is a vessel used for burning ceremonial incense. In China, incense is burned during festivals and processions to honor ancestors and household gods.

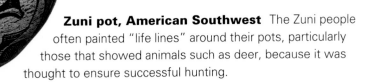

Zuni pot, American Southwest The Zuni people often painted "life lines" around their pots, particularly those that showed animals such as deer, because it was thought to ensure successful hunting.

Lending a Helping Hand

Perhaps the best thing about containers is that they're just plain useful! In fact, many people rely on them to make their work easier, particularly in jobs where it's necessary to carry lots of things.

Doctor's saddle bag, United States, 1800s

At one time, most doctors did not have offices. Instead, they made house calls to visit the sick at home. This meant they had to carry all their medical instruments and supplies with them wherever they went. This leather bag, filled with medicine bottles, is the type a doctor would carry if he were traveling on horseback.

People harvesting potatoes, Russia

Fish trap, Madagascar In Madagascar, a country east of the African continent, fishermen use fishing baskets like this, which allow fish to swim in but not out!

Blow-gun dart case, Guyana
The Arawak people of Guyana traditionally carried the darts they used for hunting in a case like this one.

Workers at tea-processing plant, Japan

Spice vendor at a local market, Turkey

Coffee-bean picker, Cameroon

Getting Carried Away

Maybe you've never thought about it this way, but suitcases and backpacks are containers, too. Because travel almost always involves carrying some of your belongings from place to place, it helps if you have an efficient way of transporting them. Nomadic people, such as the Samis of Lapland, have to carry their belongings over many many miles and for long periods of time. The Samis rely on sacks made of reindeer skins to carry their tents, blankets, food, and drink over the wintery tundra in northern Scandinavia.

Woman carrying water, India

Man carrying grain, Guinea-Bissau In West African countries, it is not unusual for people to carry goods in containers balanced on their heads.

Billum bags, Papua New Guinea These women may be carrying the bags on their backs, but they're actually supporting much of the weight with their heads! If you look closely, you'll see that the baskets have straps that wrap around the women's foreheads.

Burro baskets, Morocco The burro is a very strong animal that can tolerate heavy loads and a lot of heat. In Morocco, people often load goods into burro baskets for transport.

Carrying bananas to market, Vietnam

Straw backpacks, Nepal

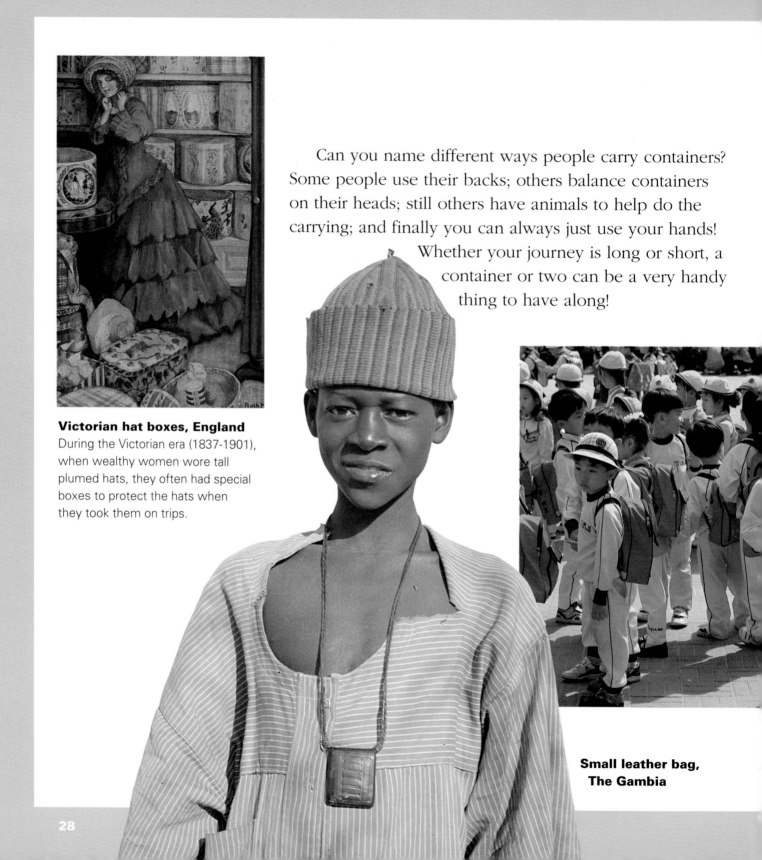

Can you name different ways people carry containers? Some people use their backs; others balance containers on their heads; still others have animals to help do the carrying; and finally you can always just use your hands!

Whether your journey is long or short, a container or two can be a very handy thing to have along!

Victorian hat boxes, England
During the Victorian era (1837-1901), when wealthy women wore tall plumed hats, they often had special boxes to protect the hats when they took them on trips.

Small leather bag, The Gambia

Farmer with basket-pack, Philippines

Airport, United States Without containers like suitcases, backpacks, and duffel bags, getting from place to place would be a lot more time-consuming and uncomfortable!

Pocketbook Pocketbooks are carried most often by women, who find them a convenient way to carry money, personal items, and other valuables from place to place.

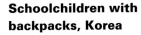

Schoolchildren with backpacks, Korea

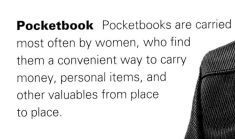

Sami reindeer sacks, Finland The Sami people, who live in far-northern Europe above the Artic Circle, spend the long winter season following their reindeer herds across miles and miles of tundra. Without the help of reindeer-skin sacks, it would be difficult to cover those kinds of distances.

Containers and the Future

As you can see, containers can be wonderful things. They're useful, often nice to look at, and some even have important cultural and religious signficance. But what about all the containers that get thrown in the trash, like milk cartons, cans, and bottles, as well as Styrofoam and cardboard boxes? These types of containers may not be as special as some of the others, but they are the most prevalent kind in the world today—and they are taking up a lot of precious space in landfills around the world! One of the healthiest things you can do for your planet is to recycle as many containers as you can instead of throwing them out. Or, if possible, use containers that don't need to be thrown out when they're empty.

Recycling center, United States Recycling is still a growing practice. Even today, there are many cities that have no recycling programs at all. The best recyclers in the world are in Japan, where 50 percent of all waste is recycled. In the United States, the rate is less than 25 percent.

Landfill Every year, the United States disposes of over 160 million tons of waste. That's equal to the weight of 40 million elephants! Because landfills take up precious space in the environment, countries around the world have looked to recycling to help ease the garbage situation.

Vendor selling recycled bottles and cans, Ethiopia

This way, when we talk about containers, we won't need to talk about overflowing landfills, and instead can focus on how containers do a wonderful job of holding us all together!

Public container for recycling paper, Quebec, Canada

Collecting cans for recycling, Taiwan

Glossary

appreciate to be aware of the worth of (p. 14)

civilization a complex society with a stable food supply, division of labor, some form of government, and a highly developed culture (p. 14)

communal referring to an activity done with a group of people (p. 20)

convenient suited to a person's comfort or ease (p. 29)

culture the beliefs and customs of a group of people that are passed from generation to generation (p. 14)

durable sturdy, long-lasting (p. 11)

elaborate having much detail (p. 16)

environment natural surroundings (p. 8)

era a period of time (p. 28)

essential necessary (p. 13)

evaporate to turn into vapor (p. 12)

faith a system of religious belief (p. 22)

fiber a threadlike material that is a part of plant tissue (p. 8)

incense a substance that gives off a fragrant odor when burned (p. 23)

manufactured made in a factory (p. 7)

nomadic referring to people who move from place to place instead of settling in one place (p. 11)

porcelain a fine hard pottery that was first produced in China (p. 14)

prevalent common (p. 30)

rituals symbolic practices or ceremonies (p. 20)

savanna a grassy plain with few or no trees (p. 11)

serenity peacefulness (p. 21)

status position or rank (p. 19)

survival the act of remaining alive (p. 14)

technology the scientific methods and ideas used in industry and trade (p. 6)

traditional handed down from generation to generation (p. 7)

trek a slow journey (p. 12)

tundra a large, treeless, almost flat plain of the arctic regions (p. 10)

Index

About the Author

Sara Corbett is a writer who lives in Santa Fe, New Mexico. She has written four other books in the *World of Difference* series.